Rebuilt, Not Replaced

How Ethical Restoration Can Save Our Roads, Wallets, and Planet

By Vitalii Tkachenko
Licensed Automotive Engineer | Founder, The Guaranteed Best Choice
Member, NIADA & FIADA

Copyright & ISBN

ISBN: (Pending Registration)
First Edition, Printed in the United States of America

To every working family who deserves affordable, safe transportation — and to every small business owner who keeps rebuilding against the odds.

QUICK REFERENCE CONTENTS

TABLE OF CONTENTS

FOREWORD

For more than a century, America's transportation story has been told through the lens of new cars, new factories, and new technologies. Yet behind every gleaming showroom and federal rebate program lies another story — one that is rarely given the recognition it deserves: the quiet, persistent work of rebuilding and restoring the vehicles we already have.

I have had the privilege of watching Vitalii Tkachenko shine a light on this overlooked sector. As a licensed automotive engineer and founder of The Guaranteed Best Choice, he has built more than a business — he has built a movement. His advocacy is rooted in three values that should matter to us all: **sustainability, affordability, and fairness.**

What makes this book extraordinary is that it connects dots most policymakers, media outlets, and even industry leaders have overlooked. Vitalii demonstrates that professionally rebuilt vehicles are not a compromise but a cornerstone of the circular economy. He shows that small, independent dealers are not merely local entrepreneurs but vital infrastructure for working families. And he explains, with the precision of an engineer and the heart of a community leader, how restoring cars can also restore trust, opportunity, and balance in our transportation system.

This is not theory. It is practice. Vitalii and thousands like him are already reducing emissions, saving families thousands of dollars, and keeping skilled workers employed — all without a penny of federal subsidy. The data, the stories, and the solutions collected in this book are evidence of what is possible when innovation comes not from billion-dollar corporations but from determined individuals who refuse to accept waste as inevitable.

Rebuilt, Not Replaced is more than a book; it is a manifesto for change. It challenges assumptions, elevates overlooked solutions, and sets a vision for a greener, fairer, more affordable future.

Every policymaker who cares about climate, every economist who cares about mobility, and every family who cares about the next generation should read these pages.

— *Maiia Tkachenko/ NIADA & FIADA Member / Sustainability Advocate*
Hollywood, Florida
2025

PREFACE

Transportation is the quiet thread that runs through every life. It decides whether a parent makes a shift, whether a student reaches class, and whether a family gets to the doctor on time. In America, that thread is fraying.

New cars keep drifting out of reach for working families. At the same time, building vehicles is one of the most resource-intensive things we do—heavy on energy, materials, and waste before the first mile. Policymakers talk about "sustainable mobility," but the spotlight often lands only on subsidies and long timelines.

There is a practical answer already here. Every day, insurers write off cars as "total losses" not because they can't be fixed, but because spreadsheets say they aren't worth it. Skilled rebuilders have long restored many of these vehicles to safe, roadworthy condition—documented, inspected, and typically priced about 30–35% below local clean-title retail. When done right and proven with a VIN-linked packet (photos, parts, alignment, SRS/ADAS, OBD, warranty), a rebuild delivers real affordability while avoiding a large share of new manufacturing emissions.

This book makes that case with data and stories. It shows that rebuilding is engineering renewal—not a shortcut—and that sustainability can live without subsidies. It argues that fair rules for small dealers—truthful auction photos, claim clocks that start at pickup, and transparent records—are essential to lower prices for families.

I write as an engineer, a rebuilder, and an advocate who has watched thousands of cars get a second life and thousands of families get a fairer deal. If there is one message to keep, it is this: **the greenest car you can buy may already exist—if we have the courage to rebuild it right.**

— *Vitalii Tkachenko*
Florida, 2025

ABOUT THE AUTHOR

Vitalii Tkachenko grew up in a family where cars were more than machines — they were part of everyday life. As a child, he spent countless hours alongside his father, working on old Soviet engines, studying every part and every system. What began as simple curiosity quickly became a lifelong passion.

That passion led him to formal education. Vitalii first enrolled in a technical college, where he earned his qualification as a junior automotive specialist. He later went on to complete a **degree in automotive engineering**, followed by a **degree in finance**, combining his love of mechanics with an understanding of how industries and markets operate. This dual foundation — engineering precision and financial insight — would shape his future as both an innovator and an entrepreneur.

One of his earliest achievements came during a **national engineering competition in Ukraine**, where he won second place by presenting a fully hydraulic three-wheeled vehicle. Far from a student experiment, the project was an early demonstration of his ability to merge innovation with practical design. It was the first public recognition of his engineering vision — and a sign of the career to come.

After moving to the United States, Vitalii turned his skills into a business that reflected both his values and his expertise. He founded **The Guaranteed Best Choice**, a Florida-based dealership specializing in the restoration and resale of insurance-salvage and flood-damaged vehicles. Over the years, he has overseen the rebuilding of thousands of cars, developing strict standards for safety, transparency, and environmental responsibility. Under his leadership, the company has become a respected model of ethical rebuilding, serving working families and supporting a wide network of subcontractors, repair specialists, and suppliers.

Today, Vitalii is recognized not only as a **licensed automotive engineer** but also as a **pioneer in sustainable restoration**. His work proves that a professionally rebuilt vehicle can be as safe and reliable as a factory-new model, while costing families **30–35% less and reducing carbon emissions by up to 80%**. He demonstrates that rebuilding is not a shortcut but a form of engineering renewal — a practical solution that bridges affordability and environmental responsibility.

What distinguishes Vitalii's career is the way he blends **engineering excellence, small-business leadership, and innovation**. He has become a voice in national conversations about affordability and sustainability, showing how restored vehicles can provide working families with access to transportation while contributing to America's climate and economic goals.

Beyond his technical and business achievements, he is deeply committed to the principle that **mobility is not a luxury, but a right**. His mission is to make safe, reliable cars accessible to the families who need them most, while reducing unnecessary waste and preserving resources. He frames rebuilding not only as an economic service, but as a form of **environmental stewardship and social responsibility**.

With more than two decades of experience, a record of innovation, and a vision that bridges engineering, entrepreneurship, and sustainability, **Vitalii Tkachenko continues to champion the idea that the future of mobility lies not only in what we build new, but in what we rebuild with integrity.**

PART I – THE CRISIS IN CARS

CHAPTER 1 – THE PRICE BARRIER NOBODY TALKS ABOUT

For millions of Americans, a car is not a luxury. It is the bridge to employment, education, healthcare, and opportunity. Without reliable transportation, parents cannot reach work, children miss school, and families are cut off from the basics of modern life. Yet today, that bridge is collapsing under the weight of costs few families can afford.

The Rising Cost of Cars

According to Kelley Blue Book, the **average price of a new vehicle in 2024 exceeded $47,500**. Even late-model used cars now average more than **$27,000**. For families living paycheck to paycheck, those numbers are staggering. The Federal Reserve reports that nearly **40% of U.S. households cannot cover an unexpected $400 expense** without going into debt. For them, the difference between a $20,000 car and a $30,000 car is not just statistics — it is the difference between mobility and immobility.

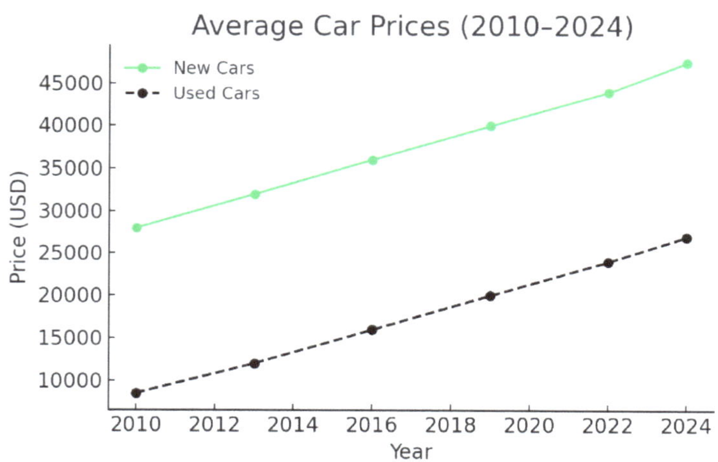

Figure 1. Average Car Prices (2010–2024).

Quick Facts: The Price Barrier

- Average new car price (2024): **$47,500+**
- Average used car price (2024): **$27,000+**
- 40% of households: cannot cover a **$400 emergency**
- Transportation = second-largest household cost after housing

Transportation Debt: A Growing Burden

Transportation debt has quietly become one of America's biggest financial burdens. A family that takes on a $30,000 car loan may end up paying over $40,000 by the time interest, fees, and insurance are added. In many cases, the monthly payment rivals what families spend on housing.

For low- and middle-income households, that means painful trade-offs: delaying healthcare, skipping childcare, or cutting back on food and essentials just to keep a car.

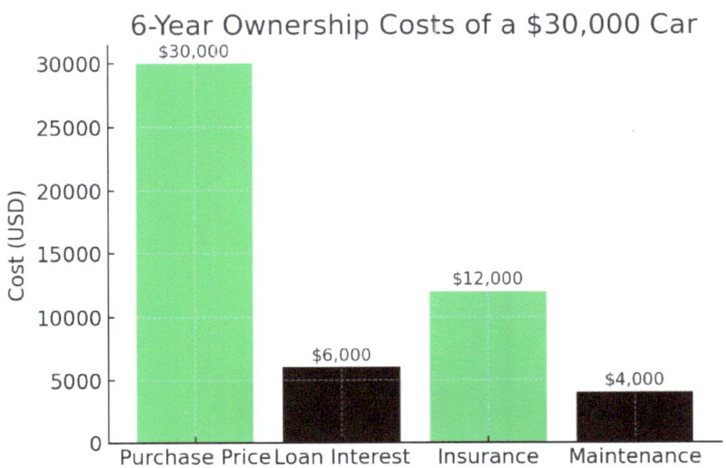

Figure 2. True Cost of Owning a $30,000 Car (6 Years).

Quick Facts: Transportation Debt

- $30,000 car loan → **$36,000+ repaid** after interest
- Insurance adds **$1,500–2,500/year**
- Real cost of a $30k car over 6 years: **$45–50k**
- Unlike mortgages or education → **car debt loses value** every month

Opportunity Lost

This is the price barrier no one talks about. It is not as visible as rising rents or health insurance premiums, but it quietly limits opportunity and deepens inequality. Without affordable mobility, families are stuck — locked out of better jobs, better schools, and better futures.

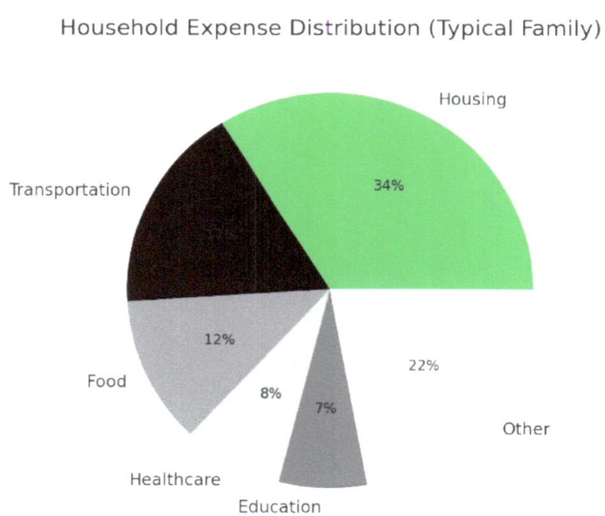

Figure 3. Household Expense Distribution (Typical Family).

An Overlooked Solution: Certified Rebuilt Vehicles

And yet, there is an overlooked solution already operating nationwide: **certified rebuilt vehicles**.

Rebuilt vehicles are cars that insurers declared "total loss" after accidents, floods, or other events — often for economic reasons rather than safety reasons. Skilled professionals restore these vehicles, repair all damage, and submit them to state inspections. Once certified, the vehicles receive a "rebuilt" title, verifying they are roadworthy.

When done to proper standards, these cars are as safe and reliable as clean-title vehicles, but they sell for dramatically less. On average, a certified rebuilt car costs **30–35% less** than its clean-title equivalent. That translates into **$8,000–$12,000 in savings** for a typical family purchase. Lower loan amounts also mean lower monthly payments, less debt, and more financial breathing room.

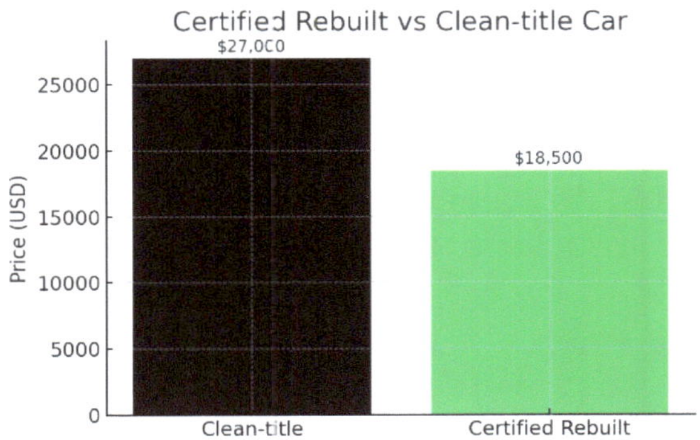

Figure 4. Price Comparison: Clean-title vs Certified Rebuilt.

Quick Facts: Certified Rebuilt Vehicles

- **30–35% less expensive** than clean-title equivalents
- Families save **$8,000–12,000** on average
- Fully inspected and state-certified
- Often come with **warranties** just like clean-title cars

Everyday Example

Consider this example: A single parent in Florida purchases a **2019 Toyota Camry** with a rebuilt title for $18,500. The same car with a clean title would cost $27,000. That $8,500 difference covers nearly a year's rent, a semester of college tuition, or several months of childcare. The rebuilt vehicle passes all state inspections, carries a warranty, and performs identically on the road.

Multiply this scenario across tens of thousands of families each year, and the national impact becomes clear. Certified rebuilt vehicles offer not just transportation, but opportunity — a pathway to stability that families otherwise could not afford.

The Policy Blind Spot

Yet despite these advantages, federal policy rarely acknowledges rebuilt vehicles. Incentives exist for new cars, especially electric vehicles, but not for certified rebuilds. As a result, working families are excluded from programs that could help them most. The silence around this issue keeps the price barrier intact.

A Different Vision

This book argues for a different vision. **Mobility should be treated as a right, not a luxury.** If reliable transportation is essential for work, education, and family life, then safe, affordable rebuilt vehicles deserve recognition, regulation, and respect.

The crisis of affordability is real, but so is the solution. Rebuilding does not just repair cars — it restores access, dignity, and opportunity. And for the families who need mobility most, that difference can change everything.

CHAPTER 2 – THE CARBON DEBT OF NEW CARS

When most people think about vehicle emissions, they picture exhaust pipes. They imagine smoke trailing behind a car in traffic, or the clean silence of an electric vehicle gliding down the road. What rarely enters the conversation is the invisible weight every car carries before it even turns its first mile: the **carbon debt of manufacturing**.

The Hidden Footprint

Building a new car is one of the most resource-intensive industrial processes on the planet. According to the International Energy Agency, manufacturing a single passenger car consumes over **30,000 kilowatt-hours of energy** — enough to power the average American home for more than two years.

The European Environment Agency estimates that producing a conventional gasoline car generates **7–10 metric tons of CO_2** before the car even leaves the factory. For electric vehicles, battery production adds even more: **an additional 2–5 tons of CO_2**, depending on the source of raw materials and the energy mix of the factory.

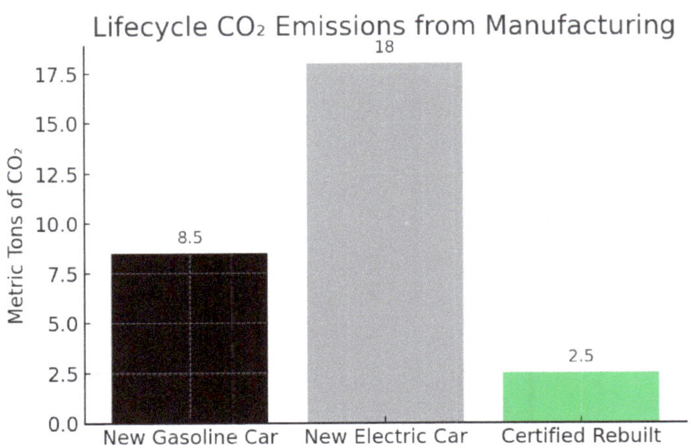

Figure 1 – Lifecycle CO_2 Emissions (Gasoline, EV, Rebuilt)

These emissions are "front-loaded." They happen before the vehicle is even driven, long before it delivers any promised fuel savings or environmental benefits. Every new car rolls out of the factory carrying an environmental debt it must spend years repaying.

Quick Facts: The Carbon Debt of New Cars

- New gasoline car: **7–10 tons CO_2** before driving.
- New electric vehicle: **17–20 tons CO_2**, with batteries accounting for up to 5 tons.
- Average U.S. household: **30,000 kWh** of electricity in 2 years = 1 car built.
- A car starts its life with a bill — the question is how long it takes to repay.

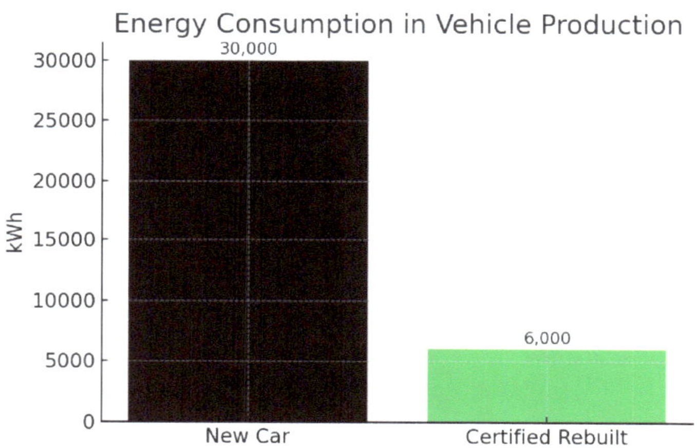

Figure 2 – Energy Consumption in Vehicle Production

Mining the Planet for Mobility

Behind every gleaming showroom model is a chain of extraction:

- **Steel and aluminum** mined and smelted with enormous energy inputs.

- **Copper and plastics** refined and shaped into wires and interiors.
- **Lithium, cobalt, and nickel** ripped from the earth for electric vehicle batteries, often in conditions that scar landscapes and displace communities.

Each of these steps leaves scars — deforestation, water contamination, toxic byproducts. The cost is not just in carbon but in ecosystems and human lives.

Rebuilding Preserves Embedded Energy

Now compare this to the process of rebuilding. When we restore an existing vehicle, we preserve most of the **embedded energy** — the energy already invested in producing the body, the frame, and the majority of mechanical systems.

Instead of starting from raw materials, rebuilding replaces only the damaged or unsafe components. When a vehicle is scrapped instead of restored, an average of nearly 2 tons of material ends up as waste — much of it unrecycled steel, plastics, and glass. By contrast, certified rebuilding diverts almost all of that material, generating only a fraction of waste. This makes rebuilding not only a carbon-saving strategy, but also a powerful way to reduce landfill input.

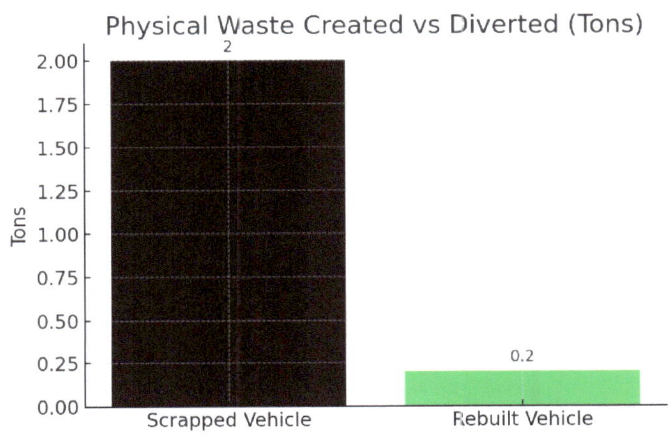

Figure 3 – Physical Waste Created vs Diverted

Studies on remanufacturing show that restoration requires **less than 25% of the energy** needed to build a comparable new vehicle.

That translates into dramatic carbon savings:

- A professional rebuild emits **2–3 metric tons of CO_2**, compared to 7–10 tons for a new gas car.
- Compared to a new EV, rebuilding can cut emissions by more than **80%**.

Quick Facts: Rebuild vs. New

- Certified rebuild: **2–3 tons CO_2.**
- New gasoline car: **7–10 tons CO_2.**
- New EV: **17–20 tons CO_2.**
- Rebuild saves **up to 80%** of emissions.
- Reuse is often greener than replace.

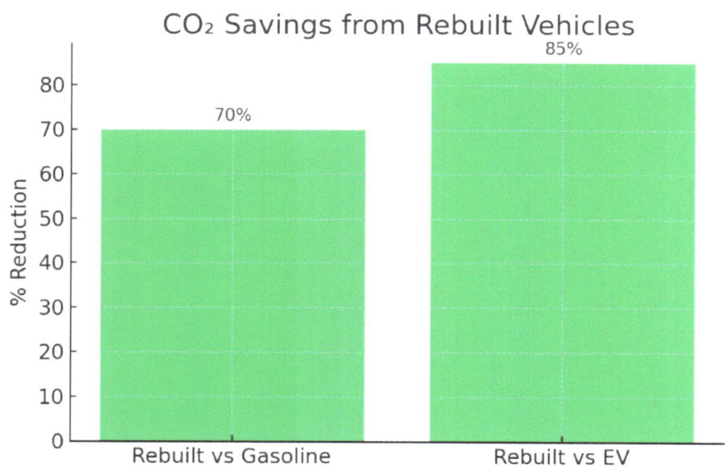

Figure 4 – CO_2 Savings from Rebuilt Vehicles

The Double Standard in Green Policy

Yet despite these numbers, federal sustainability policy rarely mentions rebuilt vehicles. Billions in subsidies go to EV manufacturing, charging infrastructure, and new fleet purchases. But not a dollar of incentive supports the simple, immediate act of keeping existing vehicles in service through certified restoration.

This oversight creates a distorted picture of what “green transportation” looks like. Policymakers celebrate factories building new electric cars while ignoring workshops that cut more emissions per vehicle by extending the life of what already exists.

Everyday Examples

Think of it this way:

- Buying a new electric SUV may save emissions over its driving lifetime — but it still starts life with a **17–20 ton carbon bill**.
- Rebuilding a five-year-old sedan may keep 80% of its parts in service and return it to the road with just **2–3 tons of added emissions**.

Which option helps the climate faster? Which is more affordable to working families? Which keeps skilled tradespeople employed in local communities rather than sending billions to global manufacturing giants?

Quick Facts: Everyday Comparison

- New EV: 17–20 tons upfront debt.
- Rebuilt sedan: 2–3 tons emissions.
- Savings: 15+ tons CO_2 = the equivalent of planting **600 trees**.

A Question of Priorities

This is not about opposing electric vehicles. They are an important part of long-term decarbonization. But while we wait for the infrastructure, supply chains, and affordability of EVs to reach scale, we cannot afford to ignore the immediate solution in front of us.

Rebuilding is not glamorous. It doesn't produce glossy marketing campaigns or futuristic headlines. But it delivers real results: lower costs for families, lower emissions for the planet, and more jobs for local economies.

Toward a Balanced Conversation

If we are serious about sustainability, the conversation must expand. We must acknowledge that the greenest car for many families is not the one fresh from a factory but the one already sitting in a driveway, a repair shop, or even a salvage yard — waiting to be rebuilt.

This is the untold story of carbon debt. Every new car carries it. Every rebuilt car pays it back.

In the next chapter, we will look at another side of the crisis: how the very marketplaces that supply vehicles — the giant auction platforms — are stacked against the small dealers who provide affordable, sustainable cars to families. Without fairness in that system, affordability and sustainability alike are at risk.

CHAPTER 3 – THE AUCTION TRAP: WHEN SMALL DEALERS LOSE

If affordable rebuilt vehicles are such a clear solution, why aren't they more widely available? Why do so many families never even see them on dealer lots? The answer lies in a hidden chokepoint in America's vehicle supply chain: **the digital auction platforms**.

Every day, tens of thousands of cars — including insurance totals, lease returns, and repossessions — are sold through giant auction companies such as Copart, IAAI, Manheim, and ACV Auctions. For independent dealers, these platforms are the main gateway to inventory. Without them, affordable vehicles never reach repair shops, lots, or families.

But behind the convenience of online bidding is a system stacked against small dealers. Buried in fine print, auction contracts shift nearly all risk to the buyer. Damages caused by the auction yard? Not their responsibility. Missing titles? Wait months with no recourse. Hidden defects? Too late — the sale is "as is."

For independent dealers who form the backbone of America's used car economy, this is not a level playing field. It is a trap.

One-Sided Contracts

Auction platforms operate under **adhesion contracts** — mandatory agreements that dealers must accept to participate. These contracts are not negotiable. They include:

- **"As-Is, Where-Is" clauses** that absolve auctions of responsibility even for undisclosed damage.
- **Mandatory arbitration** that denies dealers access to courts.
- **Extremely short claim windows** — sometimes as little as 24–48 hours, even if the car is still sitting in the auction's custody.
- **Pre-possession immunity** — auctions not liable for forklift damage or theft that occurs before the dealer takes the vehicle.

The result is simple: auctions carry little risk, while small dealers carry nearly all of it.

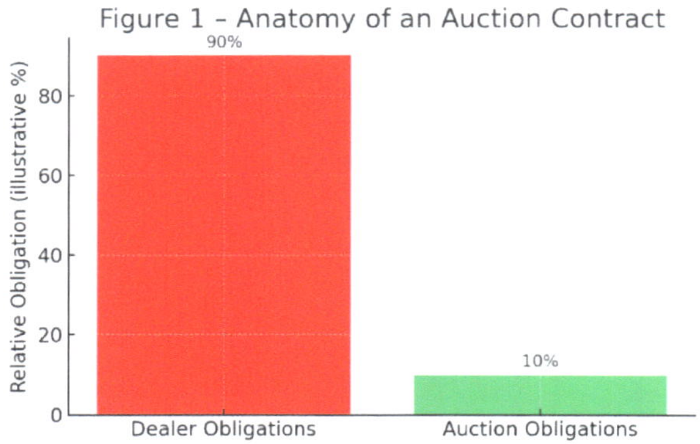

Figure 1 – Anatomy of an Auction Contract

Quick Facts: Auction Contracts

- 90%+ of salvage vehicles sold through online auctions.
- Standard terms: *as-is, where-is*.
- Claim windows: as little as **24–48 hours**.
- Dealers waive rights to courts → **forced arbitration only**.
- Auctions often immune to **yard damage or title delays**.

The Forklift Dent Problem

Consider this common story: a dealer buys a car listed as "minor cosmetic damage." Before pickup, the auction yard moves it with a forklift, bending the frame. When the dealer arrives, the car is unsellable without major repairs. He files a claim. The auction denies it, citing the "as-is" language. The dealer is left with thousands in unexpected losses.

For a large franchise with deep pockets, this is an inconvenience. For a small independent dealer, it can mean the difference between profit and bankruptcy.

Figure 2 – Forklift Claim Denial Example

Figure 2 – Claim Denial Example

Title Delays and Cash Flow

Another trap is the title delay problem. Auctions often fail to deliver titles within legal timeframes, sometimes delaying for months. During that period, dealers cannot legally sell the vehicles they purchased. Inventory sits idle, vehicles age and lose value, storage costs mount, and financing obligations (flooring costs) continue to accumulate. For small dealerships operating on thin margins, even a few such delays can cripple cash flow and disrupt operations.

Quick Facts: Title Delays

- Legal limit in most states: 30 days.
- Reality: delays of 90+ days are common.
- Consequences: vehicles lose value as they age, storage costs rise, financing expenses increase.
- Unsellable cars = frozen capital + shrinking dealer margins.

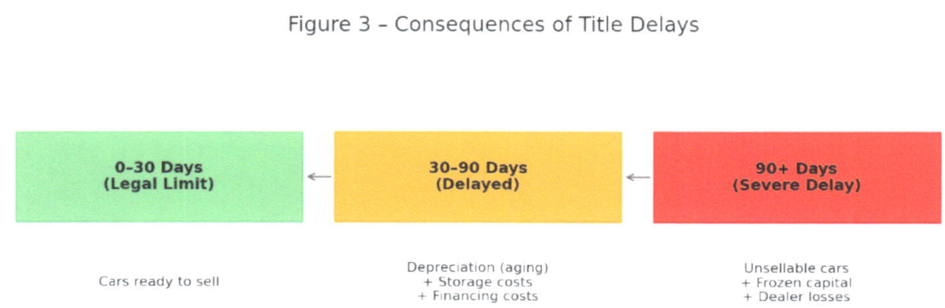

Figure 3 – Consequences of Title Delays

Digital Market Suspensions

In today's marketplace, reputation is everything. Yet independent dealers live under the constant threat of sudden account suspensions on platforms like Facebook Marketplace or CarGurus. These suspensions often occur without explanation, appeal, or recourse — wiping out years of online reputation in an instant.

For small dealers, who depend on digital visibility, such suspensions are crippling. The imbalance of power is clear: billion-dollar platforms decide who can sell, with no accountability.

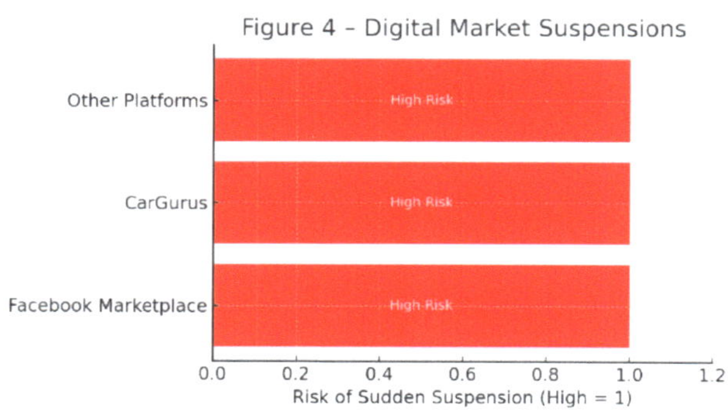

Figure 4 – Fragility of Dealer Reputation

Why This Matters for Families

The struggles of dealers may seem like an industry problem, but they directly affect families. When auctions stack the deck, small dealers cannot compete. Inventory shrinks. Prices rise. Affordable options disappear from the market.

Every unfair contract clause, every denied claim, every delayed title pushes the cost of doing business onto families in the form of higher prices and fewer choices.

In other words: **when dealers lose, working families lose too.**

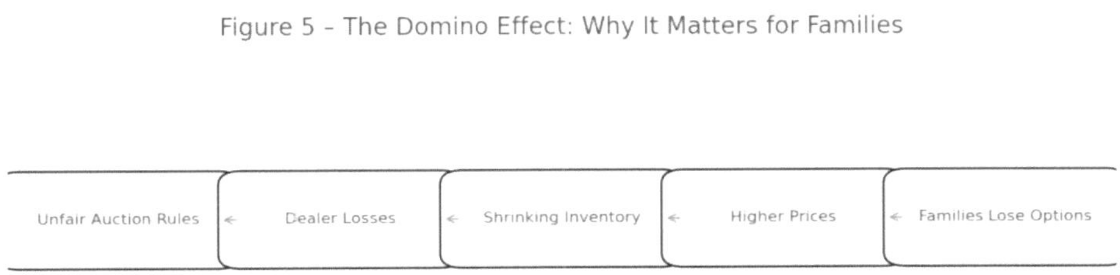

Figure 5 – The Domino Effect

Quick Facts: Why It Matters

- Independent dealers = **40,000 businesses nationwide**.
- Support **300,000+ jobs**.
- Sell cars **30–35% below franchise dealer prices**.
- Unfair auction rules → higher prices → families shut out of mobility.

The Need for Reform

If independent dealers are to continue serving communities, reform is essential. This book proposes the idea of a **Dealer Rights Bill** that would establish basic protections and level the playing field between billion-dollar auction platforms and small businesses. Such a framework would include:

- **Ban on pre-possession immunity** for auction yard damage.
- **Federal enforcement of title delivery deadlines.**
- **Fair arbitration rules** with reasonable filing windows.
- **Marketplace accountability** and clear appeal processes.
- **Transparency in digital auction listings** (recent photos, real-time damage updates).

These are not radical demands. They are simply **basic standards of fairness** needed to keep affordable mobility available for families.

Figure 6 – Proposed Dealer Rights Bill: Five Pillars of Fairness

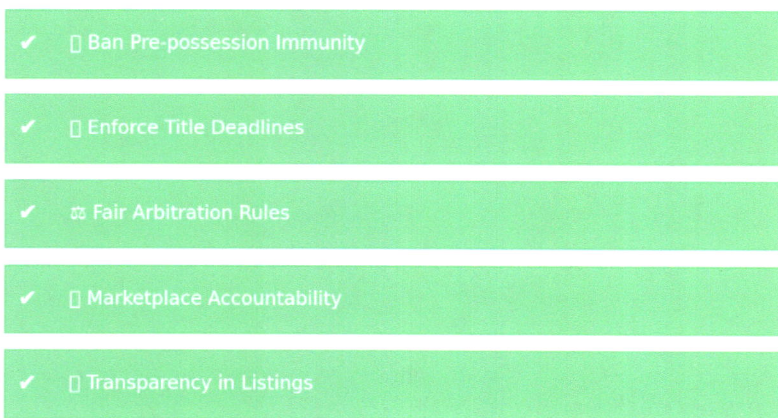

Figure 6 – Dealer Rights Bill (Five Pillars of Fairness)

Toward a Fair Marketplace

Small dealers are not asking for handouts. They are asking for rules that level the playing field, protect their businesses, and keep prices affordable for families.

If lawmakers can act to protect farmers, franchisees, and small banks from predatory contracts, they can do the same for independent dealers. Without reform, America risks losing the very businesses that provide affordable mobility to millions.

The auction trap is not just about contracts. It is about fairness, access, and the survival of an industry that families depend on every day.

PART II – THE HIDDEN SOLUTION

CHAPTER 4 – REBUILT VEHICLES: AFFORDABLE MOBILITY IN ACTION

For too long, the conversation about cars in America has been limited to two choices: buy new at sky-high prices or buy used at whatever the market allows. But there is a third path — one that is safer, greener, and dramatically more affordable. That path is the **certified rebuilt vehicle**.

Rebuilt vehicles are not junkyard experiments or unsafe shortcuts. They are cars that insurers once declared "total loss" for financial reasons — not because they were beyond repair. Skilled professionals restore these vehicles to roadworthy condition, and state inspections certify them safe. Once approved, the vehicles receive a "rebuilt" title, allowing families to purchase them at a fraction of the cost of their clean-title equivalents.

For working families, these cars can be the difference between being locked out of mobility and finally gaining a reliable way to get to work, school, or healthcare.

Quick Facts: Rebuilt Vehicles

- Declared "total loss" by insurers but restored to state-certified standards
- Pass inspections for safety, emissions, and reliability
- Cost 30–35% less than clean-title equivalents
- Offer $8,000–12,000 savings per family purchase
- Many come with warranties and documentation packets

What "Total Loss" Really Means

When people hear the phrase "total loss," they often imagine a mangled wreck. The reality is different. An insurer can declare a car a total loss simply because repair costs exceed a set percentage of its market value.

- A minor front-end collision may bend fenders and crack airbags. Replacing them at retail labor rates might cost $10,000 — more than the car's book value.
- A flood that touches interior wiring may scare insurers, even if 80% of the vehicle is untouched.
- Cosmetic hail damage can lead to a write-off, even when safety and performance remain intact.

In other words: "total loss" often means **financial shorthand**, not mechanical death. With proper rebuilding, these cars return to the road safely — and at lower cost to families.

Families Who Benefit

Consider the example of a single parent in Florida who purchases a 2019 Toyota Camry with a rebuilt title for $18,500. The clean-title version of the same car sells for $27,000. That $8,500 difference covers nearly a year's rent or multiple months of childcare. The vehicle passes all inspections, carries a warranty, and performs identically to its retail counterpart.

Another example: a delivery driver in Ohio purchases a rebuilt minivan for $14,000 instead of $21,000. Lower payments allow him to keep working, supporting his family without defaulting on other bills.

Multiply these stories by tens of thousands, and the impact becomes national. Rebuilt vehicles create **affordable mobility at scale**.

Quick Facts: Family Impact

- Camry example: $8,500 saved = a year's rent or semester tuition
- Average rebuild savings: $8,000–12,000 per vehicle
- Lower payments free up money for childcare, healthcare, or housing
- Access to cars expands access to jobs, schools, and stability

Trust Through Documentation

Skepticism is natural. Families worry: *Is a rebuilt car safe? Will it last?*

This is why the best rebuilders provide **VIN-linked documentation packets**. These include:

- Before and after photos of damage and repair
- Parts invoices, showing OEM or certified replacements
- Alignment reports proving structural integrity
- Airbag and safety system certification
- State inspection records
- Warranty details

Transparency builds trust. When families can see proof instead of promises, stigma fades and confidence grows.

Why Policy Matters

Despite clear benefits, federal and state policies rarely support rebuilt vehicles. EV buyers receive tax credits. New-car buyers benefit from subsidies and financing. But families choosing the more affordable, greener option of certified rebuilds are left without recognition.

By excluding rebuilt vehicles from incentive programs, policymakers ignore an immediate way to expand mobility while cutting emissions. Including them in rebate programs, loan access, and fleet purchases would give working families the boost they deserve.

Quick Facts: Policy Gap

- EV buyers: up to $7,500 tax credit
- Rebuilt buyers: $0 support, despite affordability + emission cuts
- No federal recognition for certified rebuilt cars
- Result: families who need help most are shut out of incentives

Mobility in Action

Rebuilt vehicles are not just an industry niche. They are a **practical mobility strategy**. They help families afford reliable cars, reduce debt, and access opportunity. They keep skilled tradespeople employed, support local economies, and extend the life of valuable machines that would otherwise be scrapped.

If we believe mobility is a right, not a luxury, then certified rebuilt vehicles must be part of the conversation.

In the next chapter, we will look at the environmental side of this solution: why rebuilt vehicles don't just save money, but also cut emissions, conserve resources, and deliver sustainability — all without taxpayer subsidies.

CHAPTER 5 – SUSTAINABILITY WITHOUT SUBSIDIES

When people talk about "green cars," the images are predictable: electric vehicles lined up at charging stations, tax credits promised by Congress, and sleek advertisements showing drivers gliding silently into the future. But sustainability is not always futuristic. In fact, one of the **greenest solutions already exists today** — and it operates without a single dollar of taxpayer subsidy.

That solution is the **independent rebuilding industry**.

Across the United States, thousands of small businesses quietly restore vehicles that insurers once wrote off as "total losses." By reusing most of the original materials, cutting waste, and extending the useful life of existing cars, these rebuilders reduce emissions on a scale that rivals — and often surpasses — new electric vehicle adoption. And unlike EVs, which depend on billions in tax credits and infrastructure spending, rebuilders do it with their own capital, skills, and networks.

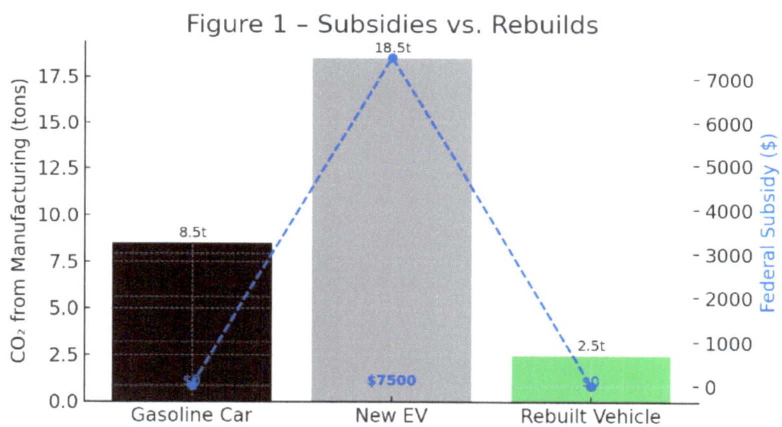

Figure 1 – Subsidies vs. Rebuilds

Quick Facts: Subsidies vs. Rebuilds

- Federal EV tax credit: up to $7,500 per vehicle

- Federal support for rebuilt vehicles: $0
- New gasoline car manufacturing: 7–10 tons CO_2
- New EV manufacturing: 17–20 tons CO_2
- Rebuilding: just 2–3 tons CO_2, no subsidies required

The Problem with One-Track Green Policy

The U.S. conversation about sustainability is dominated by EV adoption. While EVs are important for long-term decarbonization, they require:

- Massive upfront investment in factories and infrastructure
- A stable global supply chain for lithium, cobalt, and rare earth minerals
- Decades of adoption before their full impact is realized

Meanwhile, more than 800,000 insurance-totaled vehicles enter the salvage stream each year. Another 400,000 flood-damaged vehicles appear after major storm seasons. Without rebuilding, many of these are scrapped unnecessarily — despite being restorable. Each wasted vehicle represents a **lost environmental and economic opportunity**.

Sustainability by Reuse

Unlike EV programs that start from scratch, rebuilding preserves most of what has already been manufactured. On average, 60–90% of original components are kept in service, from body panels to drivetrain systems. Only the damaged or unsafe parts are replaced.

This practice avoids the massive carbon cost of mining, smelting, and fabricating new steel, aluminum, plastics, and batteries. It also keeps millions of pounds of material out of landfills every year.

From a carbon perspective, the math is striking:

- Rebuilding a vehicle emits 2–3 tons of CO_2
- Manufacturing a new gasoline car emits 7–10 tons
- Manufacturing a new EV emits 17–20 tons, mostly from the battery

That means rebuilding cuts emissions by up to 80–90% compared to new production.

Quick Facts: Reuse vs. New Production

- 60–90% of vehicle materials preserved in rebuilds
- Each rebuilt car = millions of pounds of waste avoided
- Rebuilding saves 80–90% CO_2 vs. building new
- Environmental impact: immediate, local, and scalable

Small Businesses, Big Impact

What makes rebuilding even more powerful is **who drives it**. Unlike EV programs, which rely on billion-dollar corporations and federal grants, the rebuilding industry is made up of **independent small businesses**.

These businesses:

- Invest their own money in buying and repairing vehicles
- Employ skilled technicians, inspectors, and transporters
- Support local parts suppliers and machine shops
- Sell vehicles that cost families 30–35% less than clean-title equivalents

The impact spreads outward: every dollar spent on a rebuilt vehicle circulates through local economies — not overseas supply chains.

The Policy Blind Spot

Despite clear environmental and economic benefits, rebuilt vehicles remain excluded from the green transportation conversation. They receive no incentives, no federal recognition, and no public education campaigns. Instead, policy continues to define "green" as "new and electric."

This blind spot has real consequences:

- Families who could afford rebuilt cars are left unaware or skeptical
- Dealers who provide sustainable vehicles face legal and financial disadvantages
- Millions of tons of usable material are scrapped each year in the name of progress

Quick Facts: Policy Gap

- EVs dominate federal sustainability programs
- Rebuilt vehicles = zero policy recognition
- Millions of restorable vehicles scrapped each year
- Blind spot costs families, small businesses, and the climate

Why It Matters Now

America cannot afford to wait decades for EV infrastructure to fully mature. Every year, the transportation sector remains one of the largest contributors to carbon emissions. The **fastest emissions reductions come from reusing what already exists** — not only in theory but in practice, today.

Independent rebuilders are already proving this model works: cleaner, cheaper, and immediate. They are practicing circular economy principles every day, without subsidies, and serving the families most in need of affordable mobility.

A Call to Recognize Real Sustainability

If we want a truly sustainable transportation policy, it must include rebuilt vehicles. That means:

- Recognizing certified rebuilds in federal and state green programs
- Educating consumers about safety standards and warranties
- Incentivizing lenders and insurers to support rebuilt purchases

Sustainability should not be measured only in futuristic vehicles or tax credits. It should also be measured in the quiet, practical act of reusing, restoring, and keeping families moving.

Rebuilt vehicles are proof that **sustainability without subsidies is possible** — and it is already happening at scale.

In the next chapter, we will look at how trust is built in this industry: the systems of documentation, inspection, and certification that prove rebuilt vehicles are not just affordable and green, but also safe.

CHAPTER 6 – PROOF, NOT PROMISES

For many families, the words "salvage" or "rebuilt" still sound risky. They imagine cars patched together in a backyard garage, unsafe for the road. This perception, though outdated, is not entirely unfounded. In some places, weak regulations have allowed poor-quality repairs. A few bad actors gave the entire industry a reputation problem.

The solution to this stigma is not to abandon rebuilding — it is to **raise the standards and prove the work**.

When done by professionals, rebuilt vehicles are not about shortcuts or guesswork. They are about engineering renewal, verified by testing, and documented step by step. The difference between fear and trust lies in one thing: **proof**.

What Proof Looks Like

The best rebuilders don't just repair a car — they **document the process**. This transparency transforms consumer confidence. A **VIN-linked packet** can travel with the vehicle, showing exactly what was done, when, and by whom.

A proper packet includes:

- Before-and-after photos of the vehicle
- A parts ledger (OEM, recycled, or certified new parts)
- Structural alignment and frame reports
- Airbag and safety system replacement records
- ADAS (advanced driver-assistance systems) calibration certificates
- Emissions and inspection reports
- Warranty documentation

When a family holds this packet in their hands, they are not buying a mystery. They are buying proof.

Quick Facts: Documentation Packets

- VIN-linked and tied directly to each vehicle
- Contain before-and-after photos of repairs
- Include parts invoices and safety certifications
- Show inspection results and warranty terms
- Provide transparency lenders and insurers can trust

Why Proof Matters

Without documentation, stigma persists. Families wonder if corners were cut, if safety systems really work, if the car will pass its next inspection. With documentation, those questions have clear answers.

Proof also protects lenders and insurers. When they see certified records, they can evaluate risk fairly rather than relying on blanket stigma against rebuilt titles. This expands financing and coverage options for buyers — making rebuilt vehicles even more accessible.

Inspections: The Gatekeepers of Safety

Every rebuilt vehicle must pass state inspections before being retitled. These inspections check structural integrity, safety systems, emissions, and overall roadworthiness. But standards vary from state to state.

Some states require detailed checks with photo evidence, emissions tests, and safety verification. Others allow minimal oversight. This uneven system creates confusion for consumers and fuels skepticism.

A national certification standard would solve this problem. With one set of rules for every state, families could trust that a rebuilt title means the same level of safety everywhere.

Quick Facts: Inspections

- State inspections required before a rebuilt title is issued
- Standards vary widely between states
- Strong inspections = stronger consumer trust
- National certification standard would unify expectations

The Role of Warranties

Documentation and inspections are powerful, but warranties seal the deal. A warranty proves that the rebuilder stands behind their work. Many professional rebuilders now offer 12–24 month warranties on powertrain systems, sometimes backed by national networks.

For families, this transforms perception. A car with proof and a warranty is no longer a gamble — it is a responsible purchase.

Building Industry Reputation

As more rebuilders adopt proof-based practices, the reputation of the entire industry rises. Transparency weeds out the bad actors and rewards the professionals. Families gain affordable cars. Dealers earn trust. Policymakers see a model worth supporting.

This is how stigma is overcome: not by hiding from scrutiny, but by embracing it and proving value.

Quick Facts: Why Proof Matters

- Proof reduces stigma and builds confidence
- Helps families make informed purchases
- Expands financing and insurance options

- Raises industry standards and reputation

From Stigma to Standard

The phrase "rebuilt title" does not need to trigger fear. It can — and should — signal that a vehicle was restored responsibly, inspected carefully, and documented transparently.

The key is simple: **proof, not promises**.

When proof becomes the standard, families save money, emissions are reduced, and small businesses thrive. Everyone benefits — except the myth that rebuilt cars are unsafe.

In the next chapter, we will look at the engineering side of rebuilding — the precise methods, tools, and expertise that transform damaged cars into safe, reliable vehicles once again.

CHAPTER 7 – ENGINEERING RENEWAL

A professionally rebuilt vehicle is not a compromise. Done correctly, it is the product of engineering renewal — a process that restores a car to roadworthy condition with the same precision and discipline applied in factory assembly.

The difference between a dangerous shortcut and a safe, reliable rebuild lies in the methods. Skilled rebuilders use advanced diagnostics, OEM standards, and documented procedures to ensure every system performs as designed.

For families, this means peace of mind. For the industry, it means proving that rebuilding is not second-class repair but professional engineering.

The Five Pillars of Rebuilding

A proper rebuild goes far beyond replacing parts. It follows a systematic process that touches every critical system of the vehicle:

1. **Advanced Diagnostics**
    - Computerized scanning reveals hidden electrical, electronic, and structural faults.
    - Frame lasers and borescopes detect misalignment or cracks invisible to the eye.
2. **Structural Integrity**
    - Damaged frame sections are straightened or replaced to OEM tolerances.
    - Precision alignment ensures crash energy will dissipate safely in the future.
3. **System Overhaul**
    - Brakes, suspension, steering, and drivetrain components are inspected and renewed as needed.
    - Every safety-critical component must meet manufacturer standards.
4. **Safety System Renewal**
    - Airbags, pretensioners, and electronic modules are replaced with OEM or certified units.
    - ADAS (Advanced Driver Assistance Systems) — like lane-keeping and collision warning — are recalibrated.
5. **Environmental and Emissions Compliance**

    - Vehicles must meet or exceed EPA and state emissions standards.
    - Repairs include ensuring catalytic converters, sensors, and software operate correctly.

Quick Facts: Five Pillars of Rebuilding

- Advanced diagnostics reveal hidden damage
- Frame alignment restores crash safety
- Full system overhaul of brakes, steering, suspension, drivetrain
- Safety system renewal includes airbags and ADAS calibration
- Vehicles must meet EPA/state emissions standards

Tools of Precision

Rebuilding today is a high-tech process. Professional shops use:

- **Frame alignment lasers** to measure tolerances within millimeters.
- **Electronic calibration tools** for ADAS cameras, radar, and sensors.
- **OBD-II diagnostic systems** to confirm all modules and emissions systems are functioning.
- **Ultrasonic and borescope testing** to detect cracks in welds or internal components.

This level of precision separates ethical rebuilders from amateur patchwork. It is the difference between a safe family car and a dangerous gamble.

Case Example: A Flood-Damaged Sedan

A midsize sedan arrives at auction after water intrusion during a storm. To the untrained eye, it looks unsalvageable. But in the hands of professionals:

- The vehicle is stripped and dried in controlled conditions.

- All affected wiring and sensors are replaced.
- Modules are reprogrammed and verified.
- Airbags and pretensioners are renewed.
- The car passes emissions testing and state inspection.

The result: a car that sells for 30% less than retail but performs to factory standards — safe, reliable, and affordable for a family who otherwise could not have afforded it.

The Human Side of Engineering

Behind every rebuilt vehicle is skilled labor — engineers, technicians, welders, diagnosticians. These are not machines on an assembly line but people applying knowledge, judgment, and craft.

Rebuilding is labor-intensive, which is why it supports local jobs and strengthens communities. It is also deeply human: technicians know that the cars they restore will carry children to school, parents to work, and families to safety.

Quick Facts: Why Engineering Matters

- Rebuilding uses modern diagnostic and calibration tools
- Skilled labor ensures precision and accountability
- Supports local jobs in engineering, welding, diagnostics, parts supply
- Every car restored = one family gaining safe, affordable mobility

From Repair to Renewal

Repair fixes what is broken. Renewal goes further. It ensures that a vehicle not only returns to the road but does so at a standard of safety and performance equal to or exceeding its original condition.

That is the standard true rebuilders uphold. Not "good enough." Not "cheap fix." But renewal: cars reborn for a second life of reliable service.

Why Standards Matter

Without standards, any industry risks inconsistency. That is why engineering-based rebuilding should become the national expectation. Families should know that a rebuilt title means not just "repaired" but "renewed."

The credibility of the rebuilding movement depends on engineers and technicians holding themselves to the highest level — and on policymakers recognizing those standards as valid.

Quick Facts: Renewal vs. Repair

- Repair: fixes individual problems
- Renewal: restores systems to OEM safety/performance levels
- Renewal includes structure, safety, and emissions compliance
- Families deserve renewal, not shortcuts

Engineering Renewal as the New Standard

Every rebuilt car is a second chance — not only for the vehicle but for the family who depends on it. Through diagnostics, precision, and professional renewal, we can ensure that rebuilt vehicles are not second-class but first-choice options for affordable, sustainable mobility.

In the next section of this book, we will shift from the engineering side to the **playbooks**: practical guides for buyers, dealers, and policymakers to make rebuilt vehicles work in real life.

PART III – THE PLAYBOOKS

CHAPTER 8 – BUYER PLAYBOOK: HOW FAMILIES SAVE THOUSANDS

For many families, the idea of buying a rebuilt car feels intimidating. They may wonder: *How do I know it's safe? How do I make sure I'm not getting scammed?* These are valid concerns — and the key to overcoming them is knowledge.

The truth is, families can save **$8,000–12,000 on average** by choosing a certified rebuilt vehicle over a clean-title equivalent. That savings can cover months of rent, tuition, or childcare. But those benefits only come if the purchase is made wisely.

This chapter provides a practical **step-by-step playbook** for buyers.

Step 1: Know the Price Target

Rebuilt vehicles typically sell for **30–35% less than clean-title equivalents**. This should be your starting benchmark.

If a dealer offers you a "rebuilt" car at only 5–10% below retail, walk away. The discount is not just a perk — it is the market adjustment for stigma and title branding. Aim for 30–35% below comparable clean-title listings in your area.

Quick Facts: Price Targets

- Certified rebuild: 30–35% below clean-title retail
- Average savings: $8,000–12,000
- Anything less than 25% discount: reconsider

Step 2: Demand Documentation

The difference between a risky rebuild and a reliable one is **proof**. Never buy a rebuilt car without a **VIN-linked documentation packet**.

A good packet should include:

- Before-and-after photos of damage and repairs
- Parts invoices (showing OEM or certified replacements)
- Alignment and frame reports
- Airbag and safety system replacement records
- State inspection certificate
- Warranty details

If a dealer cannot or will not provide this, the risk is too high.

Quick Facts: Must-Have Documentation

- Photos: before and after repairs
- Frame/alignment reports
- Airbag replacement proof
- Inspection and warranty papers

Step 3: Do a 10-Minute Check

Even without specialized training, buyers can do a **basic 10-minute daylight check**:

- Inspect gaps between panels — uneven spacing may signal hidden frame issues.
- Look for new paint that doesn't match surrounding areas.
- Turn the key: does the SRS (airbag) light come on and then turn off? If not, something may be wrong.
- Test drive: car should track straight, brakes should feel firm, steering smooth.

Quick Facts: Daylight Check

- Even gaps = good alignment
- SRS light cycle = airbags working
- Test drive = no pulling, smooth brakes
- Look for mismatched paint or overspray

Step 4: Ask About Warranty

Reputable rebuilders often provide a warranty — sometimes backed by national networks. Even a **12–24 month powertrain warranty** shows confidence in the work.

If no warranty is offered, negotiate price more aggressively or walk away.

Step 5: Compare Financing

Some lenders hesitate to finance rebuilt cars, but others will. Credit unions are often more flexible. Bring the documentation packet to your lender — it helps demonstrate the vehicle is safe.

If financing is difficult, weigh whether a smaller loan (because the car costs less) is still better than taking on massive debt for a clean-title car.

Step 6: Trust Your Instincts

Even with paperwork, numbers, and inspections, buying a car involves trust. If something feels wrong — if the seller is evasive, rushing you, or discourages independent inspection — walk away. There are always other options.

Buyer Checklist

Before signing any contract:

- Confirm price: at least 30% below retail
- Review VIN-linked packet with photos and reports
- Perform a 10-minute daylight check
- Confirm SRS light cycle and smooth test drive
- Ask about warranty coverage
- Bring packet to lender/insurer for confirmation

Quick Facts: Buyer Checklist

- Aim: 30–35% below clean-title retail
- Documentation is non-negotiable
- Test drive and daylight check required
- Warranty strongly preferred
- Trust your instincts — if unsure, walk away

The Family Advantage

Families who follow these steps unlock mobility at a fraction of the cost. A rebuilt vehicle is not just about saving money — it's about **gaining freedom without debt**. It means the student who can now afford to commute to college. The single parent who can drive to work reliably. The delivery driver who keeps his job instead of defaulting.

With knowledge, buyers can turn stigma into opportunity.

In the next chapter, we will look behind the counter — into the **Dealer Playbook** — to see how ethical rebuilders operate, what standards they follow, and why transparency is their strongest competitive advantage.

CHAPTER 9 – DEALER PLAYBOOK: STANDARDS FOR TRANSPARENCY

For independent dealers, reputation is everything. A single bad sale can damage years of trust. And in the rebuilt vehicle market, the stakes are even higher. Families already bring skepticism to the table. The only way to overcome it — and to grow sustainably as a business — is through **transparency and standards**.

This chapter lays out a playbook for dealers who want to succeed not just by selling cars, but by building trust.

The Business Case for Transparency

Some dealers fear that showing too much information — repair photos, inspection reports, detailed packets — will scare buyers away. In reality, the opposite is true. The more information buyers see, the more confident they become.

Transparency does three things for a dealer:

1. **Builds consumer trust** — proof beats promises.
2. **Differentiates from bad actors** — not every dealer provides documentation. Those who do stand out.
3. **Strengthens financing and insurance access** — lenders and insurers are more willing when data is available.

Quick Facts: Transparency Benefits

- Trust → higher close rates and repeat customers
- Differentiation → competitive advantage over "mystery" sellers
- Documentation → smoother financing/insurance approvals

The Dealer Standards

Dealers who want to thrive in this space should commit to these **five standards of transparency**:

1. **Provide Documentation Packets**
   - VIN-linked, with before/after photos, invoices, inspection reports, warranty terms.
2. **Perform and Share Daylight Checks**
   - Show customers the gaps, alignments, SRS light cycle, and drive test results.
3. **Offer Warranties**
   - Even limited warranties show confidence in workmanship.
4. **Be Clear About Title Branding**
   - Explain what "rebuilt" means, why it was branded, and what repairs were done.
5. **Embrace Independent Verification**
   - Encourage buyers to bring their own mechanic. A confident dealer has nothing to hide.

Quick Facts: Dealer Standards

- VIN-linked packet required for every retail sale
- Warranties (12–24 months powertrain recommended)
- Clear title branding explanation builds trust
- Independent inspections = credibility, not risk

Avoiding the Shortcuts

Some dealers cut corners to maximize margin: skipping documentation, refusing warranties, or hiding behind "as is" disclaimers. This short-term thinking destroys long-term success.

Buyers who feel misled tell others. Reviews spread online. Reputations collapse. In an industry where skepticism is already high, one bad sale does not just hurt a customer — it damages the entire rebuilding movement.

Ethical dealers must resist shortcuts. Long-term trust is worth more than short-term profit.

Building a Brand Around Proof

Dealers can go further by branding themselves around transparency. Some advertise: *"Every vehicle comes with a proof packet and warranty."* Others highlight: *"We encourage independent inspections."*

This branding flips stigma into advantage. Instead of hiding the "rebuilt" label, they proudly redefine it: not as a warning but as a guarantee of renewed safety and value.

The Ripple Effect

When ethical dealers adopt these standards, the benefits ripple outward:

- Families gain affordable mobility without fear.
- Small businesses earn repeat customers and referrals.
- Lenders and insurers expand access, reducing stigma.
- Policymakers see a credible, organized industry worth supporting.

Everyone wins — except those who profit from secrecy and shortcuts.

Quick Facts: Ripple Effect of Dealer Standards

- Families → access safe cars at lower cost
- Dealers → stronger reputation and repeat business
- Lenders/insurers → confidence in financing rebuilt cars
- Policymakers → greater recognition of rebuilding industry

The Future of Ethical Dealerships

The rebuilt vehicle industry cannot afford to be divided between ethical dealers and bad actors. The future must be one of unified standards, consistent proof, and full transparency.

Dealers who embrace this will thrive. Those who resist will fade, as consumers increasingly demand documentation, regulators raise oversight, and the market rewards integrity.

In the end, the Dealer Playbook is simple: **prove your value, protect your customers, and the business will take care of itself**.

In the next chapter, we will turn to the **Policy Playbook** — exploring how lawmakers, regulators, and industry leaders can reform the system to support rebuilt vehicles as a mainstream, sustainable, and affordable option for millions of families.

CHAPTER 10 – POLICY PLAYBOOK: REFORMING THE SYSTEM

Rebuilt vehicles provide affordability, sustainability, and access — but they remain largely invisible in U.S. policy. Federal incentives flow to new vehicles, especially EVs. State laws often vary, leaving consumers confused and dealers unprotected. Auction contracts favor billion-dollar corporations over small businesses.

If mobility is a right, then policy must reflect it. This chapter lays out a **policy playbook**: specific steps lawmakers and regulators can take to level the playing field and integrate rebuilt vehicles into the national transportation conversation.

Policy Reform #1: Federal Recognition

The first step is recognition. Today, federal programs treat rebuilt vehicles as if they don't exist. Buyers of EVs receive tax credits of up to $7,500. Fleets can access low-interest financing for new vehicles. But families who choose rebuilt cars — often the most affordable and greenest option — receive nothing.

Congress and federal agencies should:

- Include certified rebuilt vehicles in rebate and incentive programs.
- Allow nonprofits and state agencies to purchase certified rebuilds for mobility programs.
- Recognize rebuilt vehicles as part of the U.S. circular economy.

Quick Facts: Federal Recognition

- EV buyers: up to $7,500 tax credit
- Rebuilt buyers: $0 incentives
- Recognition would expand affordability + reduce emissions
- Fleets could lower costs by using certified rebuilds

Policy Reform #2: National Certification Standard

One of the biggest challenges in the rebuilt industry is inconsistency. A "rebuilt" title in one state may mean rigorous inspection and documentation; in another, it may mean the bare minimum. This inconsistency fuels consumer skepticism and lender hesitation.

Solution: create a **national restoration certification standard**.

Modeled after the National Motor Vehicle Title Information System (NMVTIS), this standard would:

- Require photo documentation of repairs.
- Link all certification to VIN history.
- Mandate emissions, safety, and structural integrity testing.
- Make certification records publicly accessible.

Quick Facts: National Certification

- Today: 50 states, 50 different standards
- Proposed: 1 national protocol for all rebuilt vehicles
- Benefits: consumer confidence, lender trust, higher resale values

Policy Reform #3: Auction Fairness

As discussed earlier, digital auctions dominate the supply chain. Yet their contracts are notoriously one-sided. Small dealers carry all the risk, while auctions claim immunity.

Reform should include:

- Ban on pre-possession immunity — auctions should be liable for damage that occurs on their property.
- Mandatory fair arbitration with at least a 7-day claim window.
- Transparency in listings with date-stamped photos and real-time updates.
- Enforcement of timely title delivery.

Quick Facts: Auction Reform

- 90%+ of salvage cars sold via online auctions
- Typical claim window: 24–48 hours — unrealistic
- Needed: 7 business days, photo proof, fair arbitration
- Fair contracts → lower risk → lower costs for families

Policy Reform #4: Financing Access

Even with proof and documentation, many lenders shy away from rebuilt vehicles. This locks families out of affordable financing, even when cars are safe and certified.

Reform should:

- Encourage credit unions and community banks to finance certified rebuilds.
- Offer risk-sharing guarantees for loans on VIN-certified vehicles.
- Educate lenders that documentation lowers risk.

Quick Facts: Financing Access

- Many banks refuse loans on rebuilt titles
- Families forced into cash-only deals or high-interest loans
- Risk-sharing + education = more financing options
- Expanded lending = more affordable mobility

Policy Reform #5: Public Education

The stigma around rebuilt vehicles remains strong. Families often assume "rebuilt" means unsafe. The best antidote is **public education**.

Federal and state agencies should:

- Run consumer awareness campaigns showing rebuilt cars can be safe and affordable.
- Provide resources explaining what a certified rebuilt title means.
- Highlight carbon and cost savings.

Quick Facts: Public Education

- Stigma = biggest barrier to adoption
- Consumer awareness campaigns can shift perception
- Proof + education = trust
- Informed families = stronger demand

Why Reform Matters

Policy is not just about laws — it's about lives. Without reform, rebuilt vehicles remain sidelined, families drown in debt, and emissions rise unnecessarily. With reform, the benefits are immediate: affordable mobility, reduced waste, stronger small businesses, and faster climate gains.

This is not a partisan issue. It is a practical one. Recognizing and supporting rebuilt vehicles is about fairness — giving working families access to safe, affordable transportation without excluding them from the nation's sustainability agenda.

The Policy Playbook in Summary

- **Federal recognition**: Include certified rebuilds in incentives and fleet programs.
- **National certification**: One standard for safety and documentation.
- **Auction fairness**: Level the contracts between dealers and billion-dollar platforms.
- **Financing access**: Expand lending to families buying certified rebuilds.
- **Public education**: Replace stigma with facts.

When these reforms are implemented, rebuilt vehicles will no longer be seen as second-class. They will take their rightful place as a core part of America's transportation system: affordable, sustainable, and fair.

In the next section of this book, we will look at **the movement ahead** — how independent dealers support the national economy, why fairness in auctions is essential, and how families' real stories prove that rebuilt vehicles change lives.

PART IV – THE MOVEMENT AHEAD

CHAPTER 11 – INDEPENDENT DEALERS AND THE NATIONAL ECONOMY

When people think about the U.S. economy, the spotlight often shines on Wall Street, technology giants, or major manufacturers. But beneath the headlines lies a quieter engine of mobility and opportunity: **independent automobile dealers**.

These businesses rarely make the news, yet they are critical to everyday life. They connect working families to affordable transportation, recycle millions of vehicles each year through restoration and resale, and support local economies with jobs and tax revenue.

Independent dealers are not just car sellers. They are multipliers — businesses that sustain entire networks of repair shops, parts suppliers, transporters, and financial service providers. Without them, the used car market would collapse into higher prices, fewer choices, and deeper inequity.

The Economic Backbone

According to the National Independent Automobile Dealers Association (NIADA), there are more than **40,000 independent dealers** in the United States. Collectively, they support over **300,000 direct jobs** and contribute billions in local tax revenue.

Their role is especially critical in rural and suburban areas, where independent lots often represent the only accessible option for affordable cars. For families priced out of franchise dealerships, independent dealers provide the lifeline.

Quick Facts: Independent Dealers

- 40,000+ independent dealers nationwide
- 300,000+ direct jobs supported
- Contribute billions in local and state tax revenue
- Often the only affordable option in rural/suburban areas

Local Impact

Every independent dealership acts as a hub in its community. A single lot may directly employ mechanics, sales staff, and detailers — but it also supports:

- Local auto body and paint shops
- Parts distributors and recyclers
- Finance companies and credit unions
- Insurance brokers and warranty providers
- Transporters and tow operators

When independent dealers thrive, these ecosystems thrive. When they struggle, the impact ripples outward, harming entire communities.

Affordability for Families

Independent dealers also play a unique role in affordability. On average, vehicles sold by independent dealers cost **30–40% less** than those sold at franchise dealerships. That price gap is what makes mobility possible for millions of families.

Without independents, the market would tilt toward large corporate sellers with less incentive to compete on price — leaving working- and middle-class families with fewer options and heavier debt burdens.

Quick Facts: Affordability Impact

- Vehicles typically 30–40% below franchise dealer prices
- Families save thousands per purchase
- Lower costs → lower debt → stronger financial stability

The Vulnerability Problem

Despite their importance, independent dealers operate with little legal protection. Unlike franchise dealerships — which are shielded by state franchise laws — independents face one-sided auction contracts, inconsistent title laws, and arbitrary digital marketplace suspensions.

This vulnerability is not just an industry problem. It is a national economic risk. If independent dealers are squeezed out, families lose access to affordable cars, small businesses close, and jobs disappear.

Why Protecting Dealers Protects Families

Every reform discussed earlier — fair auction contracts, national certification standards, financing access — is not only about protecting businesses. It is about protecting families.

When independent dealers are strong, families gain access to affordable, safe vehicles. When independents are weakened, families face higher costs and fewer options.

Quick Facts: Why Protection Matters

- Franchise dealers protected by state laws → independents are not
- Vulnerable to unfair contracts and delays
- If independents shrink, families lose access to affordable cars
- Dealer protections = family protections

Independent Dealers and the National Economy

Transportation is not optional. It is the infrastructure of daily life. Independent dealers ensure that mobility remains accessible, especially for those who need it most. Their contribution to jobs, taxes, and local economies is enormous — but too often overlooked.

Protecting and supporting this sector is not charity. It is sound economic policy. When independent dealers succeed, America succeeds.

In the next chapter, we will examine a key battleground for fairness: the auction system. Specifically, how forced arbitration and one-sided contracts undermine trust — and why reform is essential to restore balance in the marketplace.

CHAPTER 12 – RESTORING FAIRNESS IN AUCTIONS

Every year, millions of vehicles move through America's wholesale pipeline. Flood-damaged cars, insurance totals, lease returns, repossessions — all flow through massive digital auction platforms such as Copart, IAAI, Manheim, and ACV Auctions.

For independent dealers, these platforms are both lifeline and trap. They are the only practical gateway to inventory — but they operate under contracts that tilt the playing field sharply in favor of billion-dollar corporations.

The result is a system where **small businesses carry all the risk, while auctions carry almost none**.

The Problem With Arbitration

At first glance, arbitration sounds fair: a faster, cheaper alternative to court. But the way auctions use it makes it a weapon.

- Contracts force dealers into mandatory arbitration.
- Timelines for filing claims are unreasonably short — often **24–48 hours**.
- Arbitration panels are controlled by the auctions themselves.
- Most claims are denied under "as-is, where-is" rules.

This means if a car is misrepresented, damaged by forklifts, or missing key parts, dealers have virtually no recourse.

Quick Facts: Auction Arbitration

- Mandatory arbitration = no access to courts
- Claim windows: often just 24–48 hours
- Auctions judge their own disputes
- Result: dealers absorb nearly all losses

The Forklift Dent Example

Imagine a dealer purchases a truck listed as "minor cosmetic damage." Before pickup, the auction yard moves it with a forklift, bending the frame. When the dealer arrives, the damage is obvious.

He files a claim — only to be told:

- The sale was "as-is."
- The claim window expired before the vehicle was even released.
- Arbitration is denied.

The dealer absorbs the loss. For a large franchise group, it's an inconvenience. For a small independent, it can be devastating.

Title Delays

Another common issue is title delays. Dealers are often forced to wait **60–90 days or more** for paperwork that should legally be delivered in 30.

During that time, inventory sits unsellable. Cash flow collapses. Dealers are stuck with thousands in frozen capital.

Quick Facts: Title Delays

- Legal requirement: ~30 days in most states
- Reality: 60–90+ day delays are common
- Average small-dealer loss: $10,000–25,000 per month in tied-up inventory
- Families ultimately pay higher prices to cover these costs

Market Power and Vulnerability

Auctions know dealers cannot walk away. Without auctions, there is no inventory. This imbalance allows billion-dollar platforms to impose one-sided contracts with little fear of resistance.

Independent dealers — unlike franchise dealers — have no legal safety net. State franchise laws protect big corporate retailers. Independent dealers, who make up the bulk of the used-car economy, have none.

Why It Matters for Families

It is easy to think of these issues as "dealer problems." But every cost a dealer absorbs eventually lands on the family buying the car.

- Forklift damage → higher repair costs → higher sale prices.
- Title delays → frozen capital → higher interest passed to buyers.
- Arbitration denials → losses built into margins → families pay more.

When auctions are unfair to dealers, working families lose affordable options.

Quick Facts: Family Impact

- Higher dealer risk = higher retail prices
- Delayed titles = fewer cars available
- Unfair contracts = reduced competition → less choice
- Families pay the price for auction imbalance

Reform Solutions

Restoring fairness requires clear reforms:

1. **Ban pre-possession immunity** — auctions should be liable for damage on their property.
2. **Extend claim windows** — at least 7 business days from pickup.
3. **Independent arbitration** — not controlled by the auction itself.

4. **Enforce title timelines** — penalties for delays beyond 30 days.
5. **Listing transparency** — date-stamped photos, real-time updates when damage occurs.

Quick Facts: Reform Priorities

- Liability for yard damage
- Minimum 7-day claim window
- Neutral arbitration panels
- Enforceable title deadlines
- Transparent photo/document standards

Why Restoring Fairness Matters

This is not just about protecting businesses. It is about **protecting the supply chain that working families rely on**. Independent dealers cannot provide affordable cars if they are operating on razor-thin margins, absorbing endless losses, and fighting billion-dollar corporations alone.

By restoring fairness in auctions, policymakers ensure that independent dealers survive — and that families continue to have access to affordable, safe vehicles.

Fairness is not optional. It is the foundation of trust in any marketplace. Without it, the rebuilding industry — and the families it serves — cannot thrive.

In the next chapter, we will step away from contracts and economics to focus on people: the **real lives behind rebuilt vehicles**, and how affordable mobility changes everything.

CHAPTER 13 – LIVES, NOT JUST VINS

Every car carries a VIN — a Vehicle Identification Number. To insurers, auctions, and databases, that VIN is the car. It is a string of digits to be scanned, tracked, and filed.

But behind every VIN is a family. Behind every rebuilt title is a story of people who needed a chance — and found one through affordable mobility.

This chapter shifts from numbers and policy to lives. These are not hypotheticals. They are the real impacts of rebuilt vehicles on families across America.

The Single Parent

Maria, a single mother in Florida, needed a safe car to get her two children to school and herself to work. At a franchise dealership, the clean-title sedan she wanted cost $27,000. The monthly payments would have consumed half her income.

Instead, she found a certified rebuilt 2019 Toyota Camry for $18,500. With financing from her credit union, her monthly payment was $220 less than it would have been for the clean-title car.

That savings paid for groceries and after-school care. For Maria, the car was not just transportation. It was stability.

Quick Facts: Single Parent Case

- Clean-title price: $27,000
- Rebuilt price: $18,500
- Savings: $8,500
- Outcome: affordable payments → family stability

The Delivery Driver

Earl, a delivery driver in Ohio, relied on his minivan to make a living. When it was totaled after a hailstorm, insurance declared it a loss. A new van would have cost him $21,000 — more than he could borrow.

Instead, he bought a professionally rebuilt van for $14,000. It passed inspection, came with a warranty, and allowed him to keep working. Without it, he would have lost his job.

For Earl, rebuilding meant not just a car saved — but a livelihood preserved.

Quick Facts: Delivery Driver Case

- New/replacement cost: $21,000
- Rebuilt van price: $14,000
- Savings: $7,000
- Outcome: kept his job, avoided unemployment

The Student

Miguel, a first-generation college student, commuted 40 miles a day to his university. His old car broke down beyond repair. A clean-title compact cost $20,000 — far more than his part-time income could support.

He purchased a rebuilt vehicle for $13,500, financed through his credit union. His payments were manageable, and he continued his studies. Today, he is on track to graduate — something impossible without that rebuilt car.

For Miguel, rebuilding was not about steel and bolts. It was about a future.

Quick Facts: Student Case

- Clean-title car: $20,000
- Rebuilt car: $13,500
- Savings: $6,500
- Outcome: stayed in college, commuting daily

The Human Impact at Scale

These stories repeat thousands of times across America. Families priced out of clean-title cars find affordable rebuilt alternatives. Each purchase represents:

- Money saved — often thousands of dollars.
- Debt avoided — smaller loans, lower interest, less stress.
- Lives improved — access to work, school, and healthcare.

From a policy perspective, these numbers are impressive. From a human perspective, they are transformative.

Quick Facts: Human Impact

- Typical family savings: $8,000–12,000
- Rebuilt affordability = access to work, school, healthcare
- Every rebuilt vehicle = one family with more opportunity
- National impact: tens of thousands of families each year

Beyond the Stigma

For families like Maria, Earl, and Miguel, the label "rebuilt" is not a warning. It is a blessing. It means mobility at a price they can afford. It means safety, reliability, and dignity — without the crushing burden of unsustainable debt.

The stigma around rebuilt cars fades quickly when families experience the benefits. What remains is gratitude, stability, and hope.

Lives, Not Just VINs

Every time a car is rebuilt, it is not just a VIN restored. It is a life supported. A family given breathing room. A student able to keep moving forward.

Rebuilt vehicles remind us that transportation is not just about machines. It is about people.

Transition to Next Chapter

In the next chapter, we will look beyond individual stories to the future — exploring how technology, data, and policy can shape the next ten years of the rebuilding industry, and why its role will only grow more important in America's mobility system.

CHAPTER 14 – THE NEXT TEN YEARS

The rebuilding industry is not frozen in time. Like every sector of transportation, it is changing — shaped by technology, regulation, and consumer demand. The next decade will bring new challenges, but also powerful opportunities.

If policymakers, dealers, and innovators act wisely, the next ten years can transform rebuilding from a niche into a recognized pillar of America's transportation system.

AI and Data Transparency

Digital auctions have long struggled with accuracy. Photos may be weeks old, listings incomplete, damage undisclosed. This erodes trust and fuels conflict between dealers and auctions.

In the future, **artificial intelligence (AI)** can make condition reporting more accurate and transparent. Imagine:

- AI systems automatically analyzing photos for dents, frame bends, or hidden damage.
- Algorithms comparing auction listings with pickup-day scans, flagging discrepancies.
- Damage "scores" included with every VIN, accessible to all buyers.

This technology is already emerging. Within the next decade, AI could make every auction listing more truthful, reducing disputes and improving fairness.

Quick Facts: AI in Auctions

- AI photo analysis detects hidden damage
- Algorithms can flag mismatches between listing and pickup
- Damage scores create transparency for all buyers
- Result: fewer disputes, stronger trust

QR-Coded Certification

Documentation packets are powerful, but today they are mostly paper-based or PDFs. In the future, certification will be **digital, portable, and instant**.

Every certified rebuilt car could carry a **QR code** linked to its VIN history. Scan the code, and a buyer (or lender, or insurer) sees:

- Before-and-after repair photos
- Parts ledger and invoices
- Frame alignment and safety system records
- Inspection and emissions reports
- Warranty details

This kind of transparency would erase doubts, streamline lending, and elevate rebuilt vehicles into the mainstream.

Quick Facts: QR Certification

- QR code linked to VIN history
- Buyers, lenders, insurers access proof instantly
- Portable, digital, tamper-proof documentation
- Builds long-term trust in rebuilt titles

Circular Fleets

Corporations and governments are under pressure to meet sustainability goals. Fleets — delivery vans, rideshare vehicles, government cars — represent enormous opportunities.

Instead of replacing fleets with brand-new vehicles every cycle, agencies could adopt **certified rebuilt fleets**:

- Lower carbon footprint per unit.

- Lower acquisition costs.
- Jobs created in local rebuild shops instead of distant factories.

Circular fleets — vehicles restored, reused, and recertified — could become a national model for cutting emissions while saving money.

Quick Facts: Circular Fleets

- Fleets = thousands of vehicles each year
- Rebuilding saves carbon + cost simultaneously
- Supports local rebuild shops and jobs
- Aligns with corporate/government climate goals

Consumer Perception Shift

Perhaps the biggest change in the next decade will be cultural. Stigma around "rebuilt" titles will erode as transparency, certification, and warranties become standard.

Just as consumers now embrace certified pre-owned (CPO) vehicles, they will begin to trust certified rebuilt vehicles. The difference is simple: **proof will normalize trust**.

This perception shift will not happen overnight, but as more families share positive experiences, the term "rebuilt" will evolve from warning label to quality standard.

Policy Integration

The final piece is policy. For rebuilding to scale nationally, lawmakers must:

- Integrate certified rebuilds into rebate and incentive programs.
- Support national certification protocols.
- Educate consumers through public campaigns.
- Ensure fairness in auctions and financing.

The seeds are already planted. Over the next decade, these reforms can bring rebuilt vehicles into the same conversations as EVs, mass transit, and new manufacturing.

Quick Facts: Policy Future

- Incentives: rebuilt vehicles included in rebates
- National certification: one trusted standard
- Public campaigns: stigma → trust
- Fair auctions + financing = growth

The Next Ten Years in Perspective

Looking ahead, the path is clear:

- Technology will make rebuilding more transparent.
- Digital certification will make proof universal.
- Circular fleets will scale impact.
- Policy will finally recognize rebuilding as a green, affordable solution.
- Consumers will shift perception — seeing rebuilt titles not as risk, but as renewal.

In short: the next decade belongs to those who can prove, not just promise.

Transition to Conclusion

In the next and final chapter, we will return to the heart of this book's message: that the future of mobility, fairness, and sustainability in America does not depend solely on what we build new — but on what we rebuild with integrity.

CONCLUSION – REBUILT, NOT REPLACED

Every chapter of this book has told a piece of the same story: that America faces a mobility crisis — and that the solution is already in our hands.

We began with the **price barrier**, the quiet financial wall that keeps families from safe and reliable cars. We uncovered the **carbon debt of manufacturing**, the invisible emissions bill every new car carries before it even touches the road. We exposed the **auction trap**, where billion-dollar corporations stack the rules against small dealers and, by extension, working families.

And then we turned to the solution: **certified rebuilt vehicles**. We saw how they provide affordability without debt, sustainability without subsidies, and opportunity without exclusion. We explored the playbooks that buyers, dealers, and policymakers can follow to unlock this solution. And finally, we looked ahead to the next decade — a future where technology, transparency, and policy reform can bring rebuilding to the center of America's transportation system.

What Rebuilding Really Means

Rebuilding is not just about repairing a car. It is about **restoring access** for families, **preserving resources** for the planet, and **strengthening fairness** in our economy.

Every rebuilt car represents:

- A family who can afford mobility without drowning in debt.
- Tons of emissions avoided and materials reused.
- Jobs and income for skilled technicians, small businesses, and local suppliers.
- A cultural shift: from waste to renewal, from stigma to trust.

Quick Facts: The Rebuilt Promise

- 30–35% lower cost than clean-title vehicles
- $8,000–12,000 in average family savings
- Up to 80–90% carbon savings compared to new car manufacturing

- Tens of thousands of families helped each year

Mobility Is a Right, Not a Luxury

Transportation is not optional. It is the foundation of work, education, healthcare, and dignity. When we deny families affordable mobility, we deny them opportunity. When we recognize rebuilt vehicles, we give them a chance — not just to move, but to move forward.

Mobility must be treated as a **right**, not a privilege reserved for those who can afford new cars or qualify for subsidies. Certified rebuilding ensures that right remains within reach.

A National Call to Action

If America is serious about affordability, sustainability, and fairness, then we must:

- **Recognize rebuilt vehicles** in federal and state programs.
- **Create national certification standards** that guarantee safety and transparency.
- **Reform auctions and financing** to level the playing field for independent dealers.
- **Educate the public** to replace stigma with trust.

These are not radical demands. They are common-sense steps that align with our values of fairness, opportunity, and sustainability.

Rebuilt, Not Replaced

The future of transportation does not depend solely on what we build new. It depends on what we have the courage to rebuild — responsibly, ethically, and transparently.

The greenest car is not always the newest one. The fairest system is not the one that rewards billion-dollar corporations, but the one that protects families and small businesses. The most sustainable economy is not the one that discards, but the one that restores.

This is the heart of the rebuilding movement. This is the promise of a future where every family can afford mobility, where every technician's craft is valued, and where every car that can be renewed is given a second life.

Rebuilt, not replaced. That is how we lower costs, cut carbon, and restore fairness — one vehicle, one family, one community at a time.

EPILOGUE – A FINAL NOTE FROM THE AUTHOR

When I began this book, I thought about the thousands of vehicles I have seen pass through my hands. Some were bent, broken, or waterlogged. Many were dismissed as "total losses." To the insurance companies, they were numbers. To the auctions, they were transactions. To the government, they were invisible.

But to me, every one of those vehicles represented something more. A chance to restore value. A chance to give a family mobility. A chance to prove that "total loss" does not have to mean the end.

I grew up in a family where cars were more than machines — they were part of daily survival. Working side by side with my father on old engines, I learned that nothing was truly finished if you had the patience, skill, and faith to rebuild it. That lesson has guided me through every stage of my career, from my engineering studies to my work as a dealer, and now to my mission of advocacy.

This book is about cars, yes. But it is also about **people**. About the single mother who saves $8,000 on a rebuilt sedan and can finally breathe easier each month. About the student who drives to class in a rebuilt compact instead of dropping out. About the delivery driver who keeps his job because he found a van he could afford.

It is about fairness, dignity, and opportunity.

I wrote this book because I believe America can do better — not by discarding what is imperfect, but by **renewing what still has value**. I believe our transportation system can reflect our values of resilience, ingenuity, and justice. And I believe that small businesses and families deserve recognition, support, and respect.

If you take one message from these pages, let it be this: **the greenest, fairest, and most affordable future does not lie only in what we replace, but in what we rebuild with integrity.**

Thank you for joining me on this journey. I hope the next time you see a car with a rebuilt title, you will not see stigma — you will see possibility.

— *Vitalii Tkachenko*

ACKNOWLEDGMENTS

No book is written alone, and this one is no exception. Though my name is on the cover, every page carries the influence, encouragement, and support of many others.

First, I want to thank my family. To my wife, Maiia, whose strength and faith inspire me every day, and to my son, Michael, whose future reminds me why this work matters — you are my anchor and my motivation. Every long night of writing, every challenge in business, every difficult step forward was possible because of your love and patience.

To my father, who first taught me the value of rebuilding — not just cars, but life itself. The lessons I learned at your side, working on old engines, laid the foundation for everything I do today.

I am also grateful to my colleagues in the rebuilding industry — the technicians, welders, diagnosticians, and small dealers across America who prove every day that ethical rebuilding is possible. Your craftsmanship and persistence give families hope and mobility, and it is your stories that shaped much of this book.

A special thanks to the National Independent Automobile Dealers Association (NIADA) and the Florida Independent Automobile Dealers Association (FIADA), whose advocacy has highlighted the essential role of independent dealers in our national economy. Your work strengthens not just businesses, but the families who rely on us.

Finally, I want to thank the readers of this book — whether you are a policymaker, a dealer, a technician, or a family searching for affordable transportation. By engaging with these ideas, you become part of the movement to see rebuilding not as second-class, but as a standard of integrity, opportunity, and sustainability.

This book is for you, and because of you.

— *Vitalii Tkachenko*

APPENDICES

Appendix A – VIN-Linked Documentation Packet (Sample)

Vehicle: 2019 Toyota Camry, VIN: XXXXXXX
Title: Certified Rebuilt

Included in Packet:

- Before-and-after photos of damage and repair
- Parts ledger: OEM airbags, certified used fender, new radiator
- Frame alignment report: within 2 mm tolerance (OEM spec)
- Airbag module replacement: verified with scan tool
- ADAS calibration: lane departure & adaptive cruise re-calibrated
- Emissions test: passed state inspection, OBD readiness confirmed
- Warranty: 24-month/24,000-mile powertrain warranty

Note: This sample demonstrates the transparency buyers should expect. Every rebuilt car should include a VIN-linked packet of this detail.

Appendix B – Emissions Comparison Chart

Lifecycle CO_2 Emissions by Vehicle Type

Vehicle Type	Manufacturing Emissions	Average Lifecycle Savings from Rebuild
New Gasoline Car	7–10 tons CO_2	—
New Electric Vehicle (EV)	17–20 tons CO_2	—
Certified Rebuilt Vehicle	2–3 tons CO_2	**Up to 80–90% savings**

Sources: International Energy Agency (IEA), European Environment Agency, EPA lifecycle reports.

Appendix C – Buyer Checklist

Before You Buy a Rebuilt Vehicle:

- Confirm price: at least 30–35% below clean-title retail
- Demand documentation: VIN-linked packet with photos, reports, warranty
- Perform a 10-minute daylight check:
  - Panel gaps even
  - SRS (airbag) light cycles correctly
  - Smooth brakes and steering on test drive
- Confirm state inspection certificate
- Ask about warranty (12–24 months recommended)
- Bring documentation to your lender or insurer

Remember: Proof beats promises.

Appendix D – Dealer Standards of Transparency

Every Dealer Should Provide:

- A VIN-linked packet for every vehicle sold
- Clear explanation of title branding and repairs
- Warranty coverage on major systems
- Encouragement of independent inspections
- Consistent branding built on proof, not sales talk

Ethical dealers succeed by building long-term trust, not hiding details.

Appendix E – Policy Recommendations Summary

Federal Level:

- Include certified rebuilds in clean-vehicle rebates
- Establish a national certification standard
- Expand financing access through credit unions and SBA-backed programs

State Level:

- Enforce timely title delivery (30 days max)
- Standardize salvage/rebuild inspection protocols
- Support consumer education campaigns

Industry Level:

- Auction reform: fair contracts, extended claim windows, real-time photos
- Encourage fleet adoption of certified rebuilds
- Promote transparency and warranties across the industry

NOTES & REFERENCES

Chapter 1 – The Price Barrier Nobody Talks About

- Kelley Blue Book. *Average New-Vehicle Transaction Prices*. 2024.
- Board of Governors of the Federal Reserve System. *Report on the Economic Well-Being of U.S. Households*. 2023.
- U.S. Bureau of Economic Analysis. *Personal Consumption Expenditures by Major Type of Product*. 2023.

Chapter 2 – The Carbon Debt of New Cars

- International Energy Agency (IEA). *Global EV Outlook 2023*.
- European Environment Agency. *Monitoring CO_2 Emissions from Passenger Cars*. 2022.
- Union of Concerned Scientists. *Cleaner Cars from Cradle to Grave: How Electric Cars Beat Gasoline Cars on Lifetime Global Warming Emissions*. 2020.

Chapter 3 – The Auction Trap: When Small Dealers Lose

- National Independent Automobile Dealers Association (NIADA). *Dealer Industry Report*. 2023.
- Federal Trade Commission. *Unfair Contract Terms in Digital Marketplaces: Staff Perspective*. 2023.
- Automotive News. "Salvage Auction Growth and Dealer Complaints." 2022.

Chapter 4 – Rebuilt Vehicles: Affordable Mobility in Action

- U.S. Department of Transportation (DOT). *Household Transportation Accessibility and Equity Report*. 2022.
- NIADA. *Consumer Affordability and the Role of Independent Dealers*. 2021.
- Florida Department of Highway Safety and Motor Vehicles. *Salvage and Rebuilt Vehicle Inspection Standards*.

Chapter 5 – Sustainability Without Subsidies

- U.S. Environmental Protection Agency (EPA). *Lifecycle Greenhouse Gas Emissions from Automobile Manufacturing*. 2022.
- IEA. *Energy and Emissions in Vehicle Production*. 2021.
- Automotive Recyclers Association. *Annual Report on Vehicle Reuse and Recycling*. 2023.

Chapter 6 – Proof, Not Promises

- U.S. Department of Transportation (DOT). *State-by-State Rebuilt Title Requirements*. 2023.
- National Motor Vehicle Title Information System (NMVTIS). *Guide to Vehicle Title Branding*.
- NIADA. *Best Practices for Independent Dealers in Vehicle Documentation*. 2022.

Chapter 7 – Engineering Renewal

- Society of Automotive Engineers (SAE). *Automotive Structural Repair and Safety Standards*. 2021.
- EPA. *Emissions Standards for Light-Duty Vehicles*. 2023.
- Automotive Service Excellence (ASE). *Certification Standards for Collision and Rebuild Repair*.

Chapter 8 – Buyer Playbook: How Families Save Thousands

- Consumer Reports. *Used Car Buying Guide: Rebuilt Titles Explained*. 2023.
- National Consumer Law Center. *Automobile Sales and Consumer Protection Issues*.
- Credit Union National Association (CUNA). *Financing Options for Non-Traditional Vehicle Purchases*. 2022.

Chapter 9 – Dealer Playbook: Standards for Transparency

- NIADA. *Independent Dealer Best Practices Manual*. 2022.
- Better Business Bureau (BBB). *Consumer Complaints in the Used Vehicle Sector*. 2021.
- ASE. *Ethical Standards for Automotive Dealers and Technicians*.

Chapter 10 – Policy Playbook: Reforming the System

- U.S. Congress. *Inflation Reduction Act – Clean Vehicle Tax Credits*. 2022.
- Federal Trade Commission (FTC). *Contract Transparency in Wholesale Vehicle Auctions*.

- U.S. Small Business Administration (SBA). *Financing Small Dealers in Underserved Markets*.

Chapter 11 – Independent Dealers and the National Economy

- NIADA. *Independent Dealer Industry Report*. 2023.
- U.S. Bureau of Labor Statistics (BLS). *Employment in Motor Vehicle and Parts Dealers*.
- U.S. Census Bureau. *Small Business Economic Trends Report*.

Chapter 12 – Restoring Fairness in Auctions

- FTC. *Arbitration Clauses and Consumer/Business Protections*. 2023.
- NIADA Dealer Council. *Survey on Auction Contract Practices*. 2022.
- Automotive News. "Title Delays and Dealer Losses in Salvage Auctions."

Chapter 13 – Lives, Not Just VINs

- DOT. *Household Transportation Affordability Survey*. 2022.
- Consumer Federation of America. *The Role of Affordable Vehicles in Household Stability*.
- Pew Research Center. *Transportation and Economic Opportunity*. 2021.

Chapter 14 – The Next Ten Years

- IEA. *Artificial Intelligence and Automotive Manufacturing*. 2023.
- U.S. Department of Transportation. *Digital Vehicle Certification and Future Standards*.
- World Economic Forum. *Circular Economy and Vehicle Fleets*. 2022.

CALL TO ACTION

Rebuilding is more than repair. It is a movement for affordability, sustainability, and fairness. But for it to grow, it needs recognition, standards, and support.

This is a call to:

- **Families** — Ask for proof, not promises. Choose transparency, and discover that safe, affordable mobility is within reach.
- **Dealers** — Lead with integrity. Every VIN-linked packet, every warranty, every honest transaction builds trust not just in your lot, but in the entire industry.
- **Policymakers** — See beyond subsidies. Include certified rebuilds in incentive programs, reform auctions, and give families the choices they deserve.
- **Communities** — Value the small businesses and skilled workers who keep cars, and opportunities, alive.

The road ahead is not only about what we build new. It is about what we are willing to **rebuild with integrity**.

Let us prove that the greenest, fairest, and most affordable car is not in a factory, but already here — waiting to be renewed, trusted, and driven again.

Rebuilt, not replaced. That is how we move forward — one vehicle, one family, one community at a time.

www.ingramcontent.com/pod-product-compliance
Lightning Source LLC
LaVergne TN
LVHW071244210926
848486LV00008B/339
* 9 7 9 8 8 6 9 0 5 4 2 9 6 *